INTRODUCTION

Everyday Introduction to Geometry is filled with a lot of handy, stimulating activities students can complete to develop a sense about polygons, lines, angles, and more which will provide a fundamental foundation for them when they move into geometry. This collection of engaging activities is perfect to use to challenge students to think geometrically and to help introduce and reinforce basic geometry concepts in a developmental, motivational manner.

Easy to use, these reproducible activities will help students learn such geometry concepts as polygons; parallel, intersecting, and perpendicular lines; naming, classifying, and drawing angles and triangles; perimeter and area of polygons; radius, diameter, chord, area, and circumference of circles; surface area and volume; and much more.

Students will enjoy filling in charts, drawing figures, solving brain benders, and more as they learn the many concepts involved in geometry. An answer key is provided on pages 76–79 to make checking students' work simple.

Regardless of students' strengths or weaknesses when it comes to working with geometry concepts, this book provides the right variety of activities that will allow every student to experience success. You will be pleased as students begin grasping important geometry concepts and developing valuable geometry skills as they have fun working interesting and challenging activities.

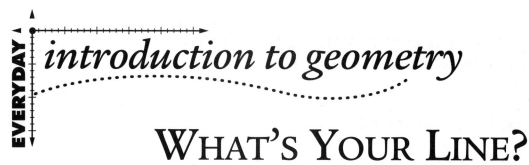

WHAT'S YOUR LINE?

A line goes on forever in both directions. To show that a line continues forever, two arrows are used on both sides of a line segment <———>. A line segment is part of a line, with a definite beginning and end. The beginning and end of a line are called endpoints. There are also points in between the endpoints of a line segment.

1. Draw points at the endpoints of the line segment below.

_____

2. Points on a line are named using capital letters. On the line segment above, name the endpoints A and C. This line segment is written as AC.

3. Using a ruler, find the exact center of the line segment. Place a point there. This is called the midpoint of the line segment. Name that point B.

4. Look at points A, B, and C on AC. Since all these points are on the same line segment, they are said to be colinear.

5. Now draw a line the same length as AC below. Write two ways that segment AC and line AC are different.

a. _____

b. _____

everyday

introduction to geometry

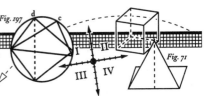

IS IT A POLYGON?

Polygons are shapes that are classified depending on their number of sides.

The word *polygon* comes from the Greek word meaning "many angles." All polygons are closed figures with many angles.

Circle the polygons below. Put an X on the figures that are not polygons. Below each figure, explain why it is or is not a polygon.

1.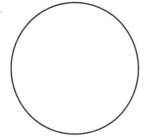

_____

_____

2.

_____

_____

3.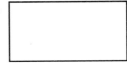

_____

_____

4.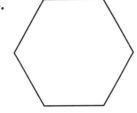

_____

_____

5.

_____

_____

6.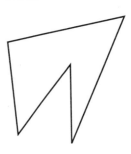

_____

_____

7.

_____

_____

8.

_____

_____

9.

_____

_____

FS-10607 Everyday Introduction to Geometry

Name_____   Date_____

PARALLEL AND INTERSECTING LINES

Parallel lines are in the same plane and do not intersect. This means that they will never cross each other no matter how far along the lines you look.

Line a is parallel to line b. This is written as a ∥ b.

1. Draw two parallel lines. Label them d and e.

2. Write "line d is parallel to line e" as shown in the above example.

When two lines meet, they are said to intersect. Lines that intersect meet at a point.

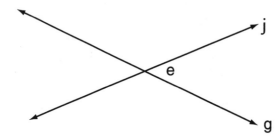

3. At what point do lines j and g intersect? _____

4. Draw two intersecting lines, i and k. Then label the point of intersection z.

INTRODUCTION TO ANGLES

An angle is formed by two rays that intersect at a common endpoint. The sides of the angle are the two rays. The vertex of the angle is the common endpoint.

The symbol ∠ means angle. The name of the angle formed by ray CA and ray CB is ∠ACB, or ∠BCA. The middle letter is always the name of the vertex.

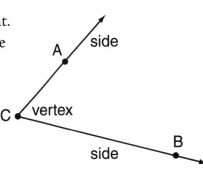

Name the angles below.

1.

_____

2.

_____

3.

_____

4.
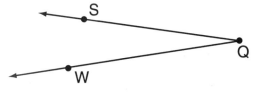

_____

5.
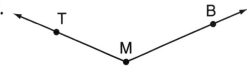

_____

6.

_____

7. **Draw and label ∠JKM.**

8. **Draw and label ∠TLA.**

9. **Draw and label ∠ABC.**

FS-10607 Everyday Introduction to Geometry

NAMING ANGLES

Name every vertex and angle for each figure below.

1.
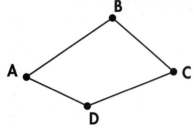

_____

_____

_____

2.
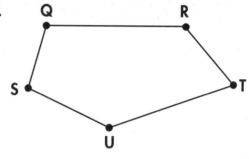

_____

_____

_____

3.
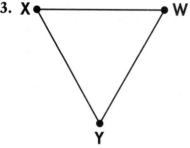

_____

_____

_____

4.
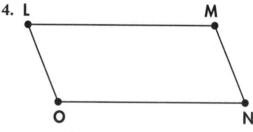

_____

_____

_____

5.
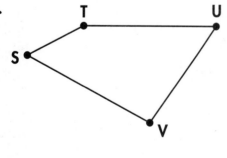

_____

_____

_____

6. **Draw and label a four-sided figure with the vertices D, E, A, and Z. Name all the angles.**

NAMING POLYGONS

Polygons are classified, or named, based on the number of sides they have. Usually the name of a polygon consists of the Greek word for the number of sides it has followed by the suffix "gon." There are two exceptions. A polygon with three sides is classified as a triangle, and a polygon with four sides is classified as a quadrilateral. Fill in the chart below. Then name the polygons.

Number of Sides	Greek Word	Name
3	—	_____
4	—	_____
5	penta	_____
6	hexa	_____
7	hepta	_____
8	octa	_____
9	nona	_____
10	deca	_____

1.

_____

2.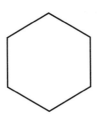

_____

3.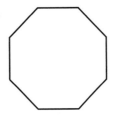

_____

4.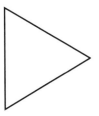

_____

5.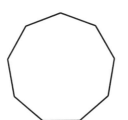

_____

6.

_____

7.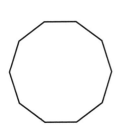

_____

8.

_____

9.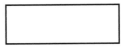

_____

FS-10607 Everyday Introduction to Geometry

NAMING QUADRILATERALS

Quadrilaterals are polygons with four sides. It is important to learn the names of some of the most common quadrilaterals.

> trapezoid—a quadrilateral with exactly one pair of parallel sides
>
> parallelogram—a quadrilateral with exactly two pairs of parallel sides
>
> rhombus—a parallelogram with all sides the same length
>
> rectangle—a parallelogram with right angles
>
> square—a rhombus with right angles; or, a rectangle with all sides the same length

1. Name the polygons below using the definitions. Be careful. Some may have more than one name.

a.
_____

b.
_____

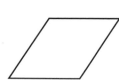
c.
_____

d.
_____

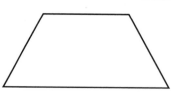
e.
_____

f.
_____

2. In the space below, draw and label an example of each of the five quadrilaterals defined above.

FS-10607 Everyday Introduction to Geometry

Name_____ Date_____

POLYGONS IN YOUR ROOM

In your classroom, there are many different geometrical shapes. Doors, windows, desks, bulletin boards, ceiling tiles, and other objects are types of polygons.

1. **In the chart below, list 20 objects in your room that are polygons. Name each object and the type of polygon it is.**

OBJECT	TYPE OF POLYGON
Example: Front door	quadrilateral, rectangle

everyday · introduction to geometry

TILE CHALLENGE

Floor tiles are often made from different types of polygons. For a floor tile installer, it is a challenge to place the correct number of floor tiles on a floor. Cut out the hexagonal "tiles" at the bottom of the page. Then place as many as you can on the rectangular "floor." Try to create a pattern with your tiles. After you finish, glue the tiles to the floor.

a. How many tiles were you able to fit on the floor without cutting? _____

b. Do you think it would be easier to fit square tiles on this floor? Explain.

_____

_____

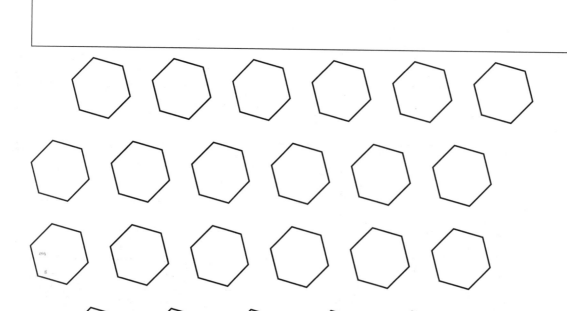

POLYGON REVIEW STACK

Name each polygon below. If it is not a polygon, cross it out. If it is a quadrilateral, be sure to include the type of quadrilateral. Next, turn to your neighbor and quiz each other!

1.
_____

2.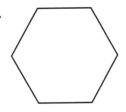
_____

3.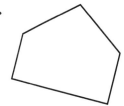
_____

4.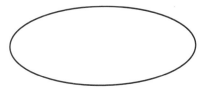
_____

5.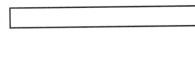
_____

6.
_____

7.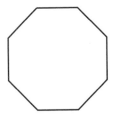
_____

8.
_____

9.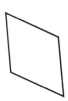
_____

10.
_____

11.
_____

12.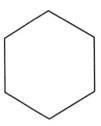
_____

13.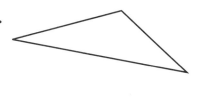
_____

14.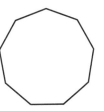
_____

15.
_____

 FS-10607 Everyday Introduction to Geometry

FINDING THE PERIMETER

Perimeter is the distance around any region. This might include the distance around your desk. The perimeter of an object is found by adding the lengths of all the sides of the object.

Find the perimeter of each figure below. Be sure to include the units of measurement in your answers.

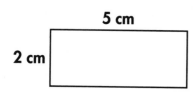

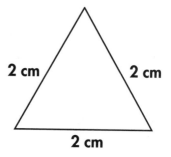

1. _____

2. _____

3. _____

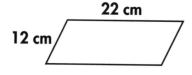

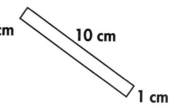

4. _____

5. _____

6. _____

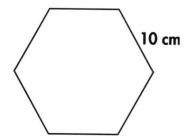

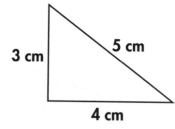

7. _____

8. _____

9. _____

FS-10607 Everyday Introduction to Geometry

MEASURING PERIMETERS OF POLYGONS

Using a ruler, measure each side of each polygon below. Write the measurement (in centimeters) next to the side. Then find the perimeters of the polygons.

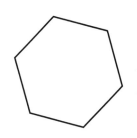

1. _____

2. _____

3. _____

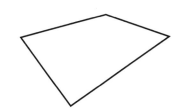

4. _____

5. _____

6. _____

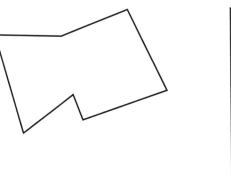

7. _____

8. _____

FS-10607 Everyday Introduction to Geometry

everyday

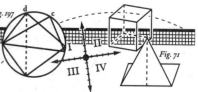

introduction to geometry

THE UNKNOWN PERIMETER

Given the perimeter, find the lengths of the unknown sides of each shape below. Show your work.

1. Perimeter = 16 m

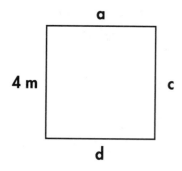

a = _____

c = _____

d = _____

2. Perimeter = 36 mm

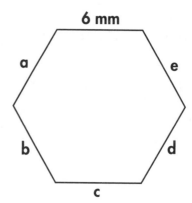

a = _____

b = _____

c = _____

d = _____

e = _____

3. Perimeter = 18 cm

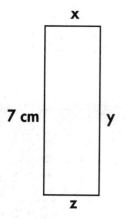

x = _____

y = _____

z = _____

4. Perimeter = 54 m

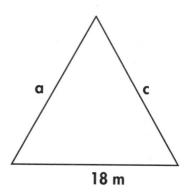

a = _____

c = _____

RADII AND CHORDS

The radius of a circle is a segment that has endpoints at the center of the circle and any point on the circle.

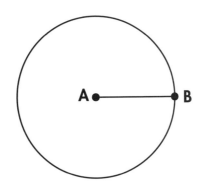

1. The center of the circle is _____ .

2. The endpoints of the radius of this circle are points _____ and _____ .

3. Using a ruler, record the length of the radius of the circle to the nearest millimeter.

---

4. On the circle, draw three more endpoints on the circle, creating radii AC, AD, AE.

A chord is a segment whose endpoints are any two points on a circle. Chords XY and MN are examples.

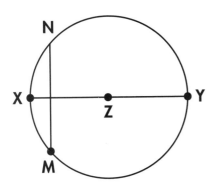

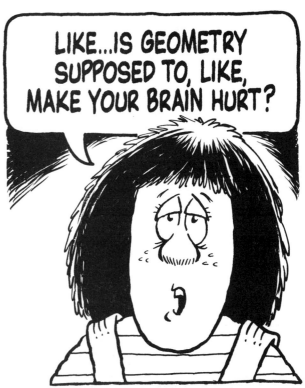

5. Draw chords TC, AB, and EF on the circle above.

6. The center of the circle is _____ .

7. Draw radius ZK on the circle.

8. Using a ruler, record the length of a radius of the circle to the nearest millimeter. _____

DIAMETER OF A CIRCLE

The diameter of a circle is a chord that passes through the center of the circle. For example, chord AE in the circle below passes through the center B. It is, therefore, the diameter of the circle.

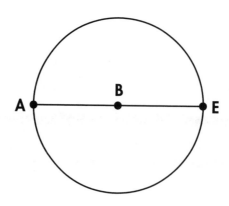

You can also see that there are two radii in the circle, AB and EB. Each radius is half the length of the diameter.

1. If the length of diameter AE is 10 cm, what is the length of radius AB? _____

EB? _____

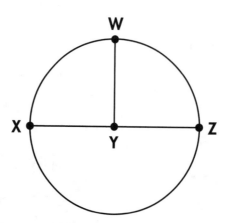

2. If the length of chord XZ is 12 cm, what are the lengths of

   a. the diameter of the circle? _____

   b. radius XY? _____

   c. radius WY? _____

3. Draw another diameter on the circle above with endpoints JK. What is the length of this diameter? _____

4. How many chords are contained in any circle? _____

everyday **introduction to geometry**

RADIUS, CHORD, AND DIAMETER REVIEW

Identify all chords, radii, and diameters below.

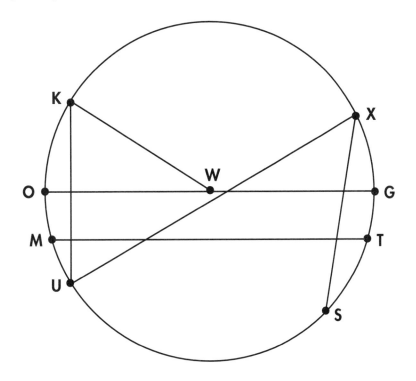

1. chords = _____

2. radii = _____

3. diameters = _____

WHAT A GREAT NATION WE LIVE IN WHERE YOUNG MEN AND WOMEN CAN SPEND THEIR TIME IDENTIFYING RADII, CHORDS AND DIAMETERS!

RIGHT ANGLES

A circle contains 360 degrees.

1. How many degrees are in $\frac{1}{4}$ of the circle below? _____

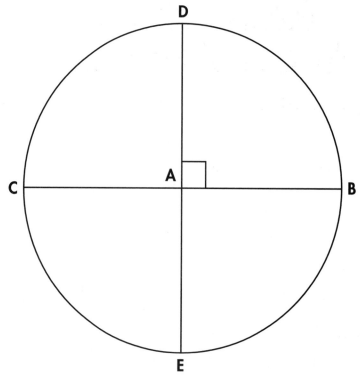

Each $\frac{1}{4}$ of the circle can be expressed as an angle.
The measure of angle DAB is 90 degrees.

All 90 degree angles are also called right angles. The lines of a right angle are perpendicular to each other. Perpendicular lines intersect to form right angles.

You can also think of a right angle as a quarter turn of a circle, or a corner in a square.

2. What is the measure of angle DAC? _____

3. How many right angles complete a circle? _____

4. Draw a square below. How many right angles are in a square? _____

5. Since all right angles are 90 degrees, how many degrees are in a square? _____

FS-10607 Everyday Introduction to Geometry

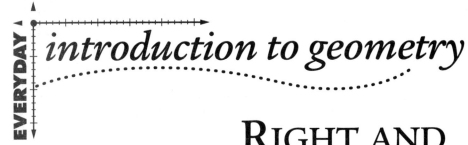

RIGHT AND ACUTE ANGLES

The point at which two perpendicular lines meet is called a right angle. A right angle has a measure of 90 degrees.

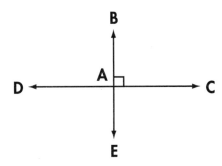

The symbol at vertex A means ∠BAC is a right angle.

1. Write the name of all other right angles in the figure above.

_____

Angles that are less than 90 degrees are called acute angles.

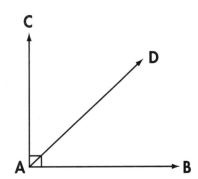

Since angle DAB is contained within a right angle, it must be less than 90 degrees and is therefore an acute angle.

2. Name another acute angle in the figure above.

_____

3. If angle DAB is equal to 35 degrees, what is the measure of the angle named in #2?

_____

everyday

introduction to geometry

ACUTE AND OBTUSE ANGLES

Obtuse angles are those which measure greater than 90 degrees.

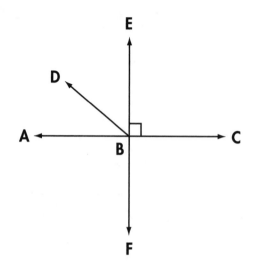

In the figure above,

1. name two obtuse angles. _____

_____

2. name two acute angles. _____

_____

3. name two right angles. _____

_____

Explore your classroom. Find two examples of each of the angles listed below. Name the object(s) that forms these angles.

4. right angle _____

_____

5. obtuse angle _____

_____

6. acute angle _____

_____

Name_____ Date_____

ANGLE MEASUREMENT

Answer the problems below without using a protractor. Show your work.

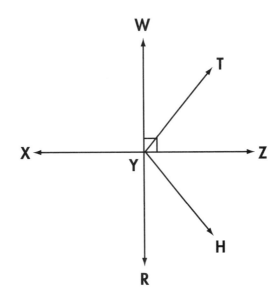

1. If the measure of angle WYT is 30 degrees, what is the measure of angle TYZ?

_____

2. If the measure of angle XYH is 120 degrees, what is the measure of angle RYH?

_____

3. If the measure of angle XYH is 120 degrees, what is the measure of angle HYZ?

_____

4. Name four right angles. _____

_____

5. Name three acute angles. _____

_____

6. Name three obtuse angles. _____

_____

7. What is the measure of angle WYZ? _____

_____

8. If the measure of angle XYT is 135 degrees, what is the measure of angle TYZ? _____

_____

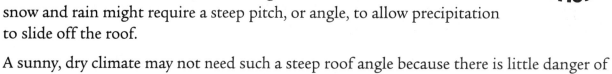

ROOF ANGLES

To build the roof of a house, certain angles are best. An area with a lot of snow and rain might require a steep pitch, or angle, to allow precipitation to slide off the roof.

A sunny, dry climate may not need such a steep roof angle because there is little danger of snow or rain pooling up and causing a leak or a collapsed roof. Examine the figure below.

1. Is this home likely to be in a hot, dry climate or a cool, wet climate? Explain your answer.

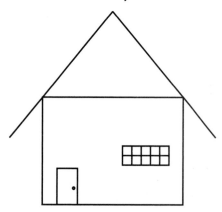

Suppose you are going to reroof the flat-roofed house below. Too steep a pitch may cost a lot of money, so keep that in mind when answering the questions below.

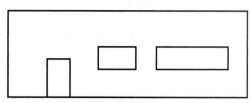

2. Draw a new roof on the house below based on the climate in your part of the world.

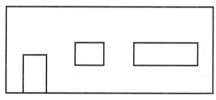

3. Explain why you chose the type of roof that you drew for your house.

_____

_____

PARALLEL AND PERPENDICULAR LINES

Parallel lines never meet. Perpendicular lines always intersect at right angles.

Lines A and B are parallel. Line C is perpendicular to both A and B.

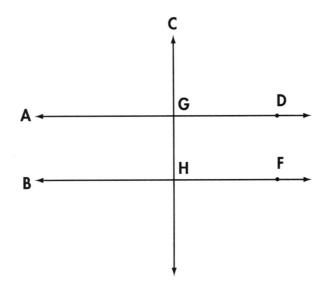

1. What is the measure of angle CGD? _____

2. What is the measure of angle GHB? _____

3. If you extend line AD and line BF forever, will they ever meet? Why or why not?

_____

_____

_____

4. Name two streets in your neighborhood that intersect at right angles.

_____

_____

CONSTRUCTING A MAP

In the city of Cloverdale, a new housing development is being built on what used to be a dairy farm.

Following the directions below, construct a map of the streets that will be built for the housing development. Be sure all streets you draw run completely from one end of the map to the other end of the map without going through any buildings. Don't forget to label the streets.

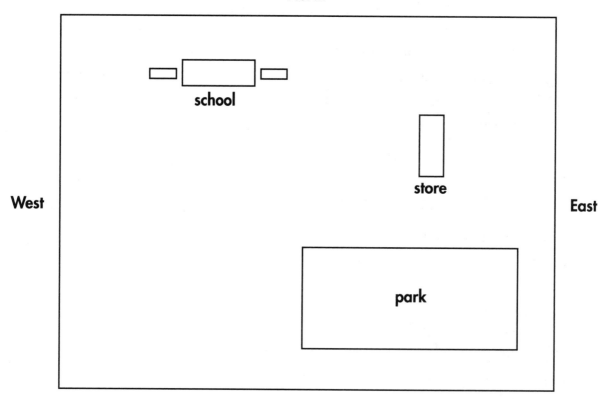

1. First Street runs from north to south along the west side of the park to the school.

2. Jackson Road is perpendicular to First Street along the north side of the park.

3. Second Street is parallel to First Street and is along the west side of the store.

4. Elm Street is perpendicular to First and Second streets and is between the north side of the store and the south side of the school.

5. Draw a stop sign at the northwest intersection of Jackson Road and First Street.

6. Draw a parking lot at the northeast intersection of Second Street and Elm Street.

introduction to geometry

FINDING THE CIRCUMFERENCE OF A CIRCLE

You already know that finding the perimeter of a polygon means adding the lengths of each side together. But how do you find the perimeter of a circle? One way is to measure it directly. Using a piece of string, measure the distance around the outside of the circle by wrapping the string around the outside of the circle. Then use a ruler to measure how much string was needed to go completely around the circle.

1. The circle is _____ cm around.

What you just measured is also called the circumference of the circle. There is another way to calculate the circumference of a circle without using string or tape. It involves the use of pi, which is equal to approximately 3.14. The symbol for pi is π. Any time you need to find the circumference of a circle, remember that $C = \pi d$.

  C means circumference.

  d means the diameter of the circle.

  π means the number 3.14.

2. Using a ruler, find the diameter of the circle above. Then calculate the circumference using the formula $C = \pi d$. Show your work.

3. If a circle has a diameter of 9 cm, then what is its circumference? _____

Name_____   Date_____

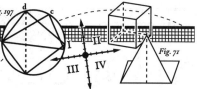
PRACTICE—CIRCLE CIRCUMFERENCE

Finding the length of the area around a circle, or circumference, means you need to use the formula for circumference (3.14 × diameter of the circle, or C = πd).

Answer the questions below. Show your work.

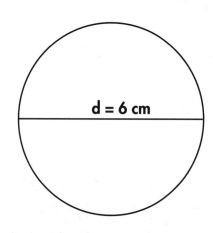

d = 6 cm

1. What is the circumference of the circle?

_____

2. What is the radius of the circle?

_____

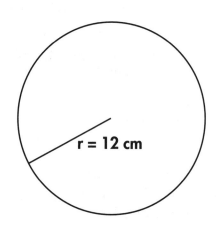

r = 12 cm

5. What is the diameter of the circle?

_____

6. What is the circumference of the circle?

_____

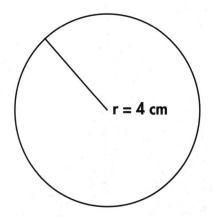

r = 4 cm

3. What is the diameter of the circle?

_____

4. What is the circumference of the circle?

_____

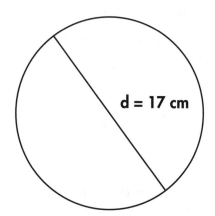

d = 17 cm

7. What is the radius of the circle?

_____

8. What is the circumference of the circle?

_____

MANIPULATIONS OF CIRCUMFERENCE

To find the circumference of a circle, the formula C = πd is used. If given the circumference, can you determine the diameter of the circle?

Example: If the circumference of a circle is 25 cm, what is its diameter?

Since C = πd, and π is always 3.14,

25 = 3.14 × d.

Now divide both sides of the equation by 3.14 to get d by itself.

$$\frac{25}{3.14} = \frac{3.14}{3.14} \times d$$

7.96 = 1 × d

d = 7.96

Answer the questions below. Show your work. Round answers to the nearest hundredth.

1. If the circumference of a circle is 50, what is the diameter of the circle?

_____

2. If C = 75, what is the diameter of the circle?

_____

3. What is the radius of the circle if C = 100?

_____

4. If the radius of a circle is 22, what is the circumference?

_____

5. If C = 85, what is the radius of the circle?

_____

6. If C = 29, what is the radius of the circle?

_____

MANIPULATIONS OF CIRCUMFERENCE?! SORRY, MAN, I'M JUST THE CARTOONIST!

AREA OF A POLYGON

The amount of surface of a region is called its area. Remember that the perimeter of a region is the length of all the sides added together. The area equals the amount of space on the entire surface of the region. Suppose the picture below is a section of a playing field.

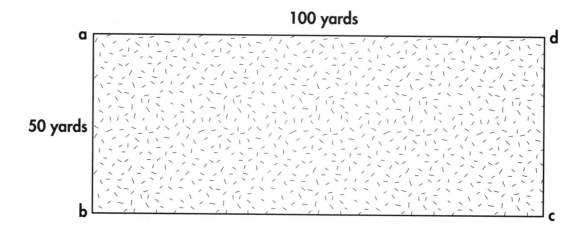

a
100 yards
d

50 yards

b
c

1. What is the perimeter of this region? _____

In the previous question about perimeter, you were asked to determine how long the lines bordering the field are. To calculate the area of the surface of the field (the shaded region), you must multiply the length of the region by the width of the region, or 100 yards × 50 yards.

2. What is the area of the field? _____

Since you are multiplying 100 yards × 50 yards in the above question, you really did this: 100 × 50 × yards × yards. What will you do with the yard units? Square them. Therefore, any time you calculate the area of a region, always remember that the units are squared and appear like this: yards^2. Make the change to your answer above if you haven't already done so.

What would the units in #2 above look like if you had measured the field in

3. inches? _____

4. meters? _____

A rectangle is 8 feet long by 6 feet wide.

5. What is the perimeter of the rectangle? _____

6. What is the area of the rectangle?_____

AREA OF A SQUARE REGION

Suppose your parents are going to buy new carpet for your house. They will put new carpet in your room and your brother Kyle's room if you can figure out how much carpet is needed for both rooms. Given the dimensions of the rooms below, answer the following questions.

1. How many square meters of carpet are needed for your room?

_____

2. How many square meters of carpet are needed for Kyle's room?

_____

3. How much carpet is needed for both rooms?

_____

4. Calculate the perimeter of both rooms together. How does this number compare to the area of the region?

_____

_____

_____

5. Why can't you use perimeter to calculate how much carpet is needed?

_____

_____

_____

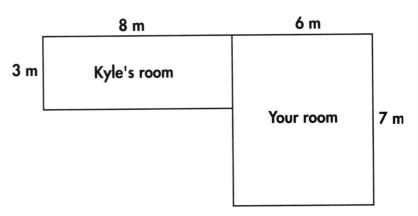

Show your work on all problems.

AREA AND PERIMETER REVIEW

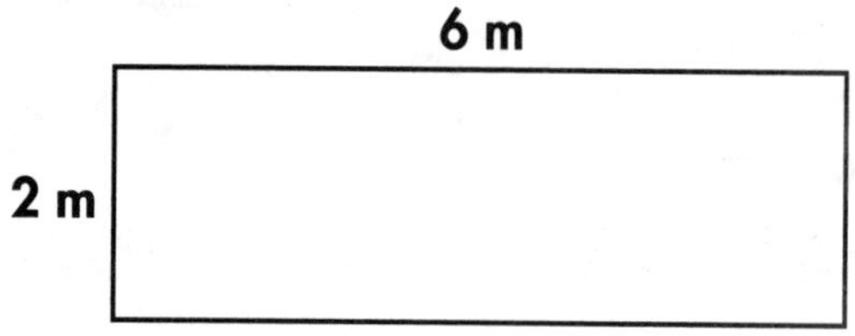

6 m

2 m

1. What is the perimeter of the figure above? _____

2. What is the area of the figure above? _____

3. If the perimeter of a square is 16 cm, what is the area of the square? Draw a picture of the square to help you.

4. Using a ruler, determine the perimeter and area of the front cover of your math textbook. Draw your book below and show all measurements.

everyday introduction to geometry

e

a

b

$6x =$

$6x - 3x = 3$

$3x = 18$

$\dfrac{3x}{3} = \dfrac{18}{3}$

$x = 6$

Fig.

Name_____ Date_____

THE MISSING LENGTH

Given the measurement of one side of a four-sided figure and the area, you can still determine the perimeter and area if you remember that parallel sides of a square or rectangle are the same length.

Example: If the area of a square is 16 cm², how long is each side? Since all 4 sides of a square are the same length, $\frac{16}{4} = 4$ cm for each side. With a rectangle, only the parallel sides are the same length. You must be given the length of one side to solve the problem. If the area of a rectangle is 16 cm², and b is 8 cm long, how long is the adjacent side h?

Since A = b × h

$16 = 8 \times h$

$\frac{16}{8} = \frac{8}{8} \times h$

$2 = h$

Answer the questions below. Show your work.

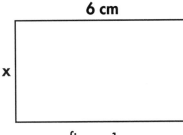

6 cm

x

figure 1

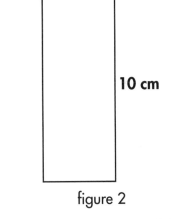

y

10 cm

figure 2

1. What is the length of side x if the area of figure 1 is 24 cm²?

2. What is the length of side y if the area of figure 2 is 40 cm²?

figure 3

3. What is the length of each side of figure 3 if the area is 48 cm²?

everyday

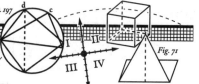

introduction to geometry

A NEW ENTERTAINMENT SYSTEM

Congratulations! You have just won a new entertainment system including a television, VCR, stereo receiver, and CD player. Now you have to buy an entertainment center to fit your entertainment system. Here are the components you won.

Television:

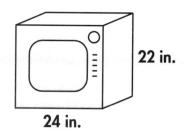

22 in.

24 in.

VCR:

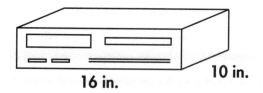

16 in.   10 in.

Stereo Receiver:

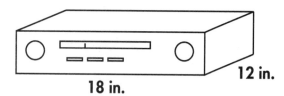

18 in.   12 in.

CD Player:

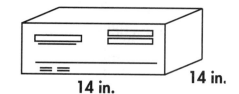

14 in.   14 in.

1. Calculate and record the area of shelf space each component will need.

Television: _____   VCR: _____

Stereo Receiver: _____   CD Player: _____

2. On the back of this page, draw an entertainment system with shelves that would fit all components. Include the measurements for each shelf.

Name _____ Date _____

 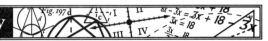
FINDING THE AREA OF A POLYGON

You can find the area of an irregular polygon by dividing it into square or rectangular regions. Look at the example below.

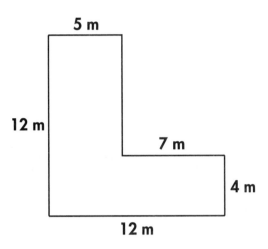

To find the area of the entire polygon, divide it into two rectangles.

Use figure 2 to answer questions 5–7.

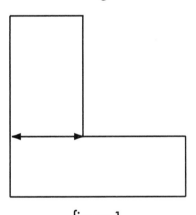

figure 1

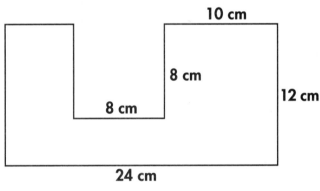

figure 2

1. Fill in the lengths of all sides in figure 1.

2. Calculate the area of the smaller rectangle.

_____

3. Calculate the area of the larger rectangle.

_____

4. What is the area of the entire region?

_____

5. Divide figure 2 into smaller rectangles and find the lengths of all the sides.

6. Find the area of each small rectangle.

_____

7. What is the area of the entire region?

_____

AREA OF A NONRECTANGULAR PARALLELOGRAM

A rectangle (figure A) and nonrectangular parallelogram (figure B) are shown below.

figure A

figure B

Finding the area of any parallelogram means multiplying the base times the height. To find the height of figure A, measure the side adjacent to the base. But how can you be sure of the height of figure B?

To find the height of figure B, you must find the length of the altitude drawn from the top to the base of the figure as shown below.

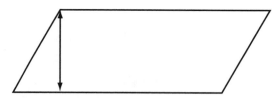

1. If the base of the above parallelogram is 10 m, and the altitude is 3 m, what is the area?

Find the areas and perimeters of the parallelograms below.

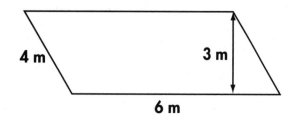

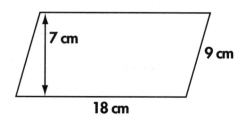

2. A = _____

P = _____

3. A = _____

P = _____

     FS-10607 Everyday Introduction to Geometry

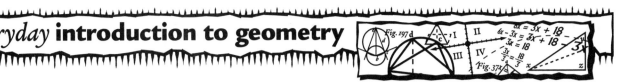

MORE PARALLELOGRAM PRACTICE

Find the area and perimeter of each parallelogram below.

1.

4 cm 3 cm 16 cm

A = _____

P = _____

2.

12 in. 11 in. 15 in.

A = _____

P = _____

3. 2 ft. 22 ft.

A = _____

P = _____

4.

17 m 21 m 36 m

A = _____

P = _____

FS-10607 Everyday Introduction to Geometry

AREA OF A TRIANGULAR REGION

1. What is the area of the rectangle below?

A = _____

4 cm

10 cm

2. If you divide the rectangle in half using a diagonal line, what do you get?

_____

What is the area of each section? _____

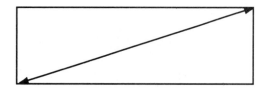

3. Since each triangle is half of the rectangle, what is the area of each triangle?

_____

You can now see that the formula to find the area of a triangle is $\frac{1}{2}$ b × h.

4. What is the area of a triangle that has a base length of 6 cm and a height of 10 cm?

_____

5. What is the area of a triangle that has a height of 2 ft. and a base length of 4 ft.?

_____

1/2 BXH?
ISN'T THAT THE
THEORY BY
ALBERT FRANKENSTEIN?

everyday introduction to geometry

introduction to geometry

EVERYDAY

AREA OF A TRIANGULAR REGION—PRACTICE

To find the area of a triangular region, you must multiply the length of the base times the height and divide by 2, or $A = \frac{1}{2}bh$.

Find the area of the triangular regions below. Show your work.

1.

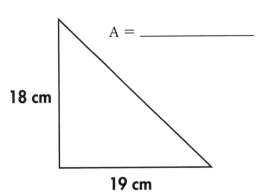

A = _____

18 cm

19 cm

4.

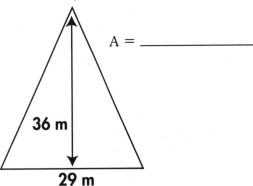

A = _____

36 m

29 m

2.

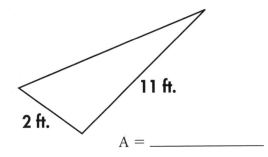

11 ft.

2 ft.

A = _____

5.

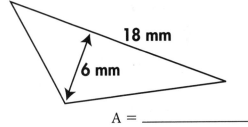

18 mm

6 mm

A = _____

3.

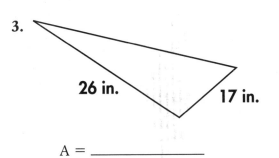

26 in.

17 in.

A = _____

HOWZABOUT A LITTLE TRIANGLE PRACTICE?

FS-10607 Everyday Introduction to Geometry

Name_____ Date_____

everyday

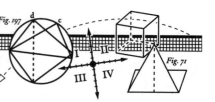

introduction to geometry

KITE FLIGHT

Flying kites can be a lot of fun. To fly well, a kite must have enough surface area to catch the wind. If it is too large and heavy, however, a kite will only be able to fly during extremely windy days. Many kites are really triangles in different forms.

Calculate the surface area of each kite below. Use the formula A $= \frac{1}{2}$bh. Show your work.

1.

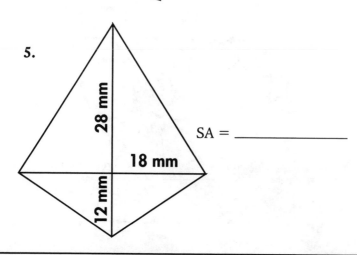

10 in.

11 in. 11 in.

17 in.

SA = _____

2.

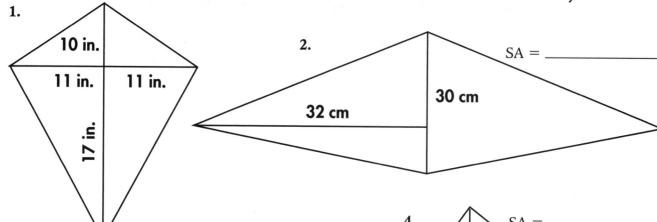

SA = _____

30 cm

32 cm

3.

31 mm

26 mm

SA = _____

4.

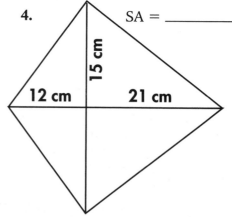

SA = _____

15 cm

12 cm 21 cm

5.

28 mm

18 mm

12 mm

SA = _____

TAKE IT FROM OL' BEN... KITE FLYING CAN BE AN ELECTRIFYING EXPERIENCE!

e v e r y d a y introduction to geometry

MATH MANIPULATIONS

Given the area and the length of one side of a triangle, you can find the length of the base or the height.

Example: Area $= 24$ cm², base length $= 4$ cm

Plug in the formula. $24 = \frac{1}{2}(4) \times h$

Simplify. $24 = 2 \times h$

Get the h by itself. $\frac{24}{2} = \frac{2}{2} \times h$

$12 = h$

Check your answer. Is $24 = \frac{1}{2} \times 4 \times 12$ true? Yes. Therefore, your answer is correct.

Find the lengths of the indicated segments.

1. A $= 16$ cm²   b = _____

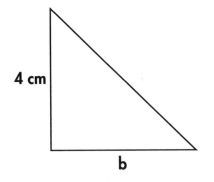

3.   A $= 90$ mm²   h = _____

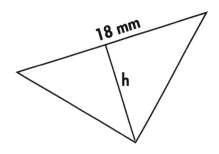

2. A $= 72$ in.²   h = _____

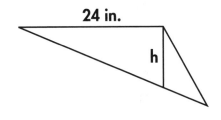

4.   A $= 72$ ft.²   b = _____

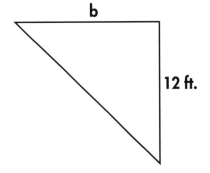

TRIANGULAR AND RECTANGULAR REGION REVIEW

1. **Using your centimeter ruler, find the area and perimeter of the rectangular region below. Then find the area of the shaded region. Show your work.**

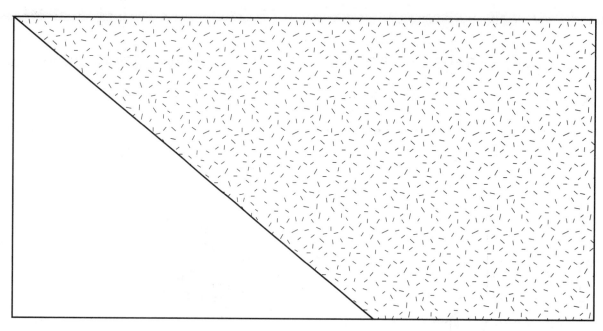

a. Area of rectangle: _____

b. Perimeter of rectangle: _____

c. Area of shaded region: _____

OKAY, LET'S REVIEW...
GEOMETRY IS THE
CURE FOR INSOMNIA.

FLOOR PLAN

Suppose you are going to have new flooring placed in your home. You will be using a combination of tile and carpet. Below is a diagram of your plan showing where you will put tile and where you will put carpet.

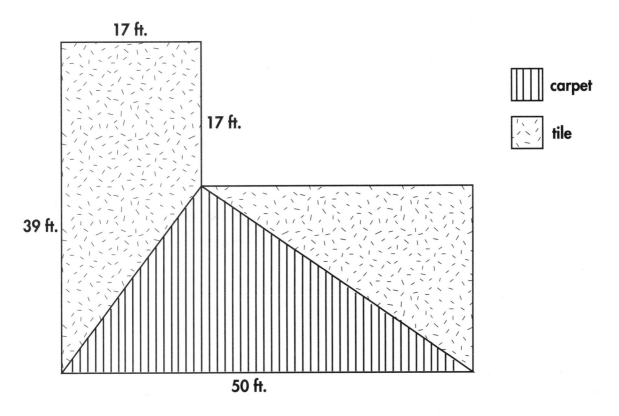

1. How many square feet of tile will you need to buy? _____

2. How many square feet of carpet will you need to buy? _____

HERE'S THAT 60,000 SQUARE FEET OF CARPET YOU ORDERED. YOU REALLY OUGHTA PAY MORE ATTENTION IN GEOMETRY CLASS.

Name_____ Date_____

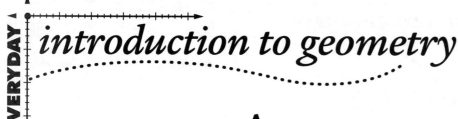

introduction to geometry

AREA OF A
CIRCULAR REGION

The perimeter, or circumference, of a circle is determined by the formula $C = \pi d$, where $\pi = 3.14$ and $d =$ diameter of the circle.

This is only the region along the outside of a circle. To find the area of the entire surface of a circle, the formula $A = \pi r^2$ is used.

Example:	Find the area of a circle with a diameter of 6 cm.
	First determine the radius. $\quad r = \frac{1}{2} \times 6 \text{ cm} = 3 \text{ cm}$
	Square the radius. $\quad 3^2 = 3 \times 3 = 9 \text{ cm}^2$
	Multiply this quantity by π. $\quad 3.14 \times 9 \text{ cm}^2 = 28.26 \text{ cm}^2$

Find the area of each circle.

1.

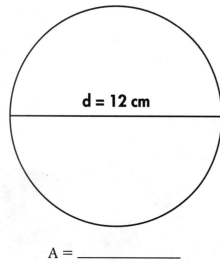

d = 12 cm

A = _____

2.

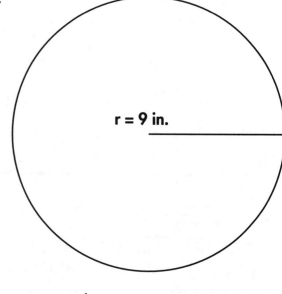

r = 9 in.

A = _____

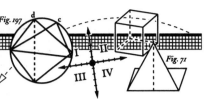

everyday

introduction to geometry

AREA OF CIRCLES

Using the formula $A = \pi r^2$, find the areas of the circles below.

Use a ruler to determine the radius of circles 1 and 2 to the nearest millimeter.

Show your work.

1.

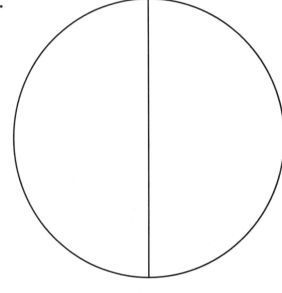

A = _____

2.

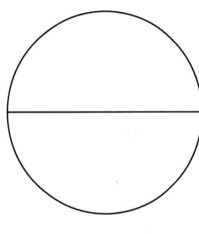

A = _____

Find the areas of the circles below.

3.

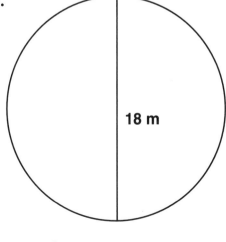

18 m

A = _____

4.

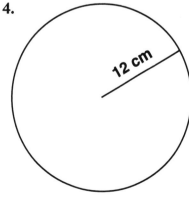

12 cm

A = _____

FS-10607 Everyday Introduction to Geometry

everyday | **introduction to geometry**

BAKERY MANIA

Mr. Badgett owns a bakery. His bakery is famous for its huckleberry pies. One day, Mr. Badgett received an order for 21 of his famous pies, and they had to be done within two hours. Mr. Badgett thought for a moment. He had enough ingredients and enough pie tins, but two hours! That meant he would only have time to bake one batch of pies. Would he have enough room for all those pies baking at once in the giant bakery oven? Help Mr. Badgett find out.

Here is one of the pie tins Mr. Badgett will use.

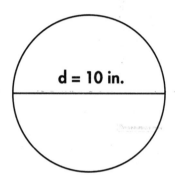

d = 10 in.

Here is the shelf in his oven.

12 ft.

IT'S HUCKLEBERRY TIME!!

8 ft.

Showing all your work, determine whether there is enough room in the oven for all 21 pies.
(**Hint:** Make sure all your units are the same.)

FS-10607 Everyday Introduction to Geometry

MORE MATH MANIPULATIONS

If the area of a circle is given, the diameter of the circle can be found using the formula $A = \pi r^2$.

Example: What is the radius of a circle with $A = 66$ cm²?

$$66 = \pi r^2$$
$$66 = 3.14 \times r^2$$
$$\frac{66}{3.14} = r^2$$
$$21.02 = r^2$$
$$\sqrt{21.02} = 4.58 = r$$

Find the radius and diameter of each circle below. Show your work.
(Round each radius to the nearest tenth.)

1. Area = 39 cm²

    radius = _____

    diameter = _____

2. Area = 86 cm²

    radius = _____

    diameter = _____

3. Area = 19 ft.²

    radius = _____

    diameter = _____

4. Area = 43 m²

    radius = _____

    diameter = _____

5. Area = 27 in.²

    radius = _____

    diameter = _____

6. Area = 55 mm²

    radius = _____

    diameter = _____

TOTAL AREA OF RECTANGULAR PRISMS

To find the total area of a prism, you must add the areas of all the faces together.

A rectangular prism has six faces—top, bottom, left, right, front, and back.

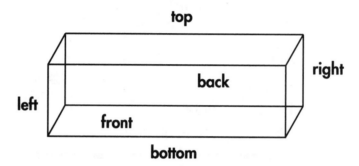

Find the total area of each rectangular prism below. Show your work.

1.

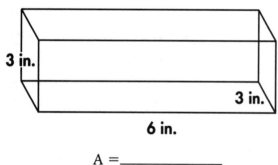

A =_____

2.

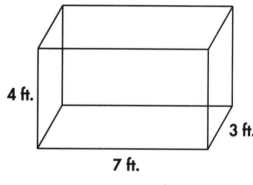

A =_____

3.

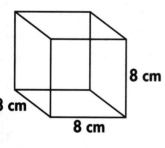

A =_____

4.

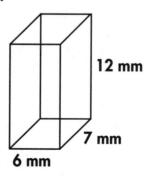

A =_____

REAL PRISMS—TOTAL SURFACE AREA

Using a ruler or meterstick, find the total surface area of each object named. Write the dimension of each face.

1. the door to your classroom

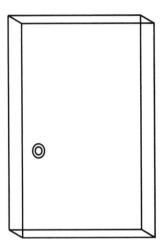

2. any book other than your math textbook

introduction to geometry

TOTAL AREA
OF A CYLINDER

The cylinder is a very common shape in the modern world. Water pipes and soda cans are just two examples of cylinders. To calculate the total area of a cylinder, you must break it up into parts that you are familiar with.

If you look carefully at this cylinder, you can see that it is really made up of two circles and one rectangle.

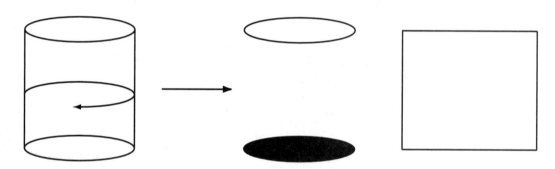

To find the total area, you need to find the area of the top and bottom of the cylinder plus the lateral area. The lateral area equals the circumference of the base times the height. Since circumference of a circle = πd, lateral area = πdh. To calculate the total area of a cylinder, the operations are πdh + (πr² × 2).

Determine the total area of each cylinder below. Show your work.

1. 2 cm

12 cm

A = _____

2. 6 mm

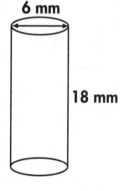

18 mm

A = _____

everyday

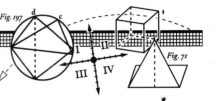

introduction to geometry

VOLUME OF SOLIDS

All solid objects take up space. The volume of an object is the amount of space taken up by it. In ancient Greece, a man named Archimedes discovered that if you drop an object into a tub of water, the water rose up. It turned out that the water rose the exact amount as the volume of the object placed in the water.

Using Archimedes' principle, write down the volume of each object placed in 100 liters of water.

1. After a cannonball is dropped in the water, the total water level is 130 liters. What is the

volume of the cannonball? _____

2. After a baby hippopotamus takes a bath in the water, the total volume of the water is 200

liters. What is the volume of the hippo? _____

3. A fishing weight is dropped into the water, and the total volume is 101.5 liters. What is the

volume of the fishing weight? _____

The amount of space taken up by a prism can be found by remembering that it exists in three-dimensional space. Those dimensions are length, height, and depth. By multiplying these dimensions together, we can arrive at the volume of a rectangular prism.

The volume of the prism to the right is equal to 50 mm × 15 mm × 24 mm = 18,000 mm³. It is important to remember that this time, we are multiplying the units (in this case, mm) together three times, so the answer is to the third power, or cubic units.

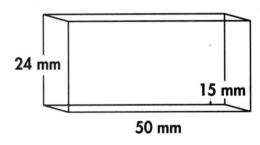

24 mm

15 mm

50 mm

Determine the volume of the rectangular prisms below. Show your work.

4. 5 cm

60 cm **6 cm**

V = _____

5.

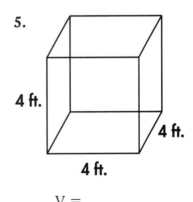

4 ft.

4 ft.

4 ft.

V = _____

Name_____ Date_____

PRISM VOLUME

In each rectangular prism below, find the total surface area and the volume. Which number is greater? Explain why for each figure.

1.

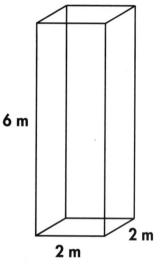

6 m

2 m

2 m

Total Surface Area = _____

Volume = _____

Greater Number = _____

Why? _____

2.

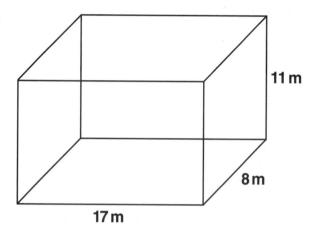

11 m

8 m

17 m

Total Surface Area = _____

Volume = _____

Greater Number = _____

Why? _____

3.

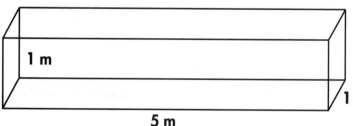

1 m

1 m

5 m

Total Surface Area = _____

Volume = _____

Greater Number = _____

Why? _____

More Volume of Solids

Using a ruler or meterstick, determine the volume of the objects listed below. Measure to the nearest cm. Draw and label a picture. Show your work.

1. your textbook

2. your desk (if your desk is not rectangular, use one that is)

3. a fish tank in your room (or any sort of animal cage)

4. a cabinet in your room

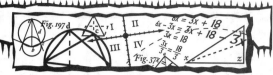

everyday **introduction to geometry**

SIMILAR FIGURES

Similar figures are shapes that are enlarged or reduced compared with the original. Look at the two figures to the right.

These figures are exactly the same shape, yet one is a smaller version of the other. They are, therefore, similar figures. Look at the next pair of figures below.

Although one figure is smaller than the other, the figures are not the same shape. They are not similar figures.

1. Write two rules that help you determine if two figures are similar.

_____

_____

Circle the similar figures in each group. Explain why the dissimilar figure does not belong.

2.

_____   _____

_____   _____

3.

4.   **5.**   **6.**

_____   _____   _____

_____   _____   _____

_____   _____   _____

SIMILAR FIGURES AND CORRESPONDING PARTS

everyday introduction to geometry

Look at the pair of similar figures below.

1. Give two reasons why these are similar figures.

_____

_____

Similar figures have identical parts that are different sizes. These parts are called corresponding parts. Looking at the figures above, name four corresponding parts.

2. _____

_____

_____

_____

Corresponding parts of figures include angles and segments. Look at the similar figures below.

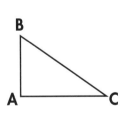 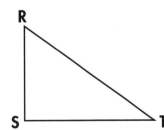

You can see that angle C corresponds to angle T. They are corresponding parts of the triangles.

3. Name the other pairs of corresponding angles.

_____

4. Name the corresponding angles of the triangles to the right.

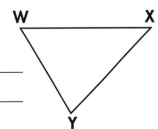 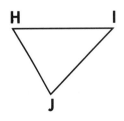

_____

_____

introduction to geometry

CORRESPONDING PARTS
OF POLYGONS

Similar figures have corresponding parts. These parts include segments and angles.

Circle the pairs of corresponding figures. Then identify the pairs of corresponding angles.

1.

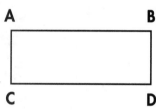
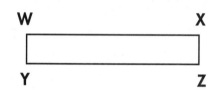

2.

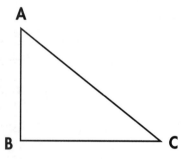

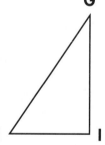

3.

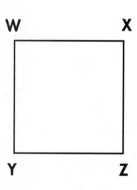

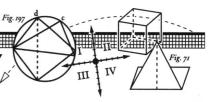

everyday

introduction to geometry

CORRESPONDING PARTS OF SIMILAR FIGURES

Look at the similar figures below. Pay careful attention to how the segments in each figure correspond.

Segments CD and FG correspond. In addition, segments DE and GH are corresponding segments.

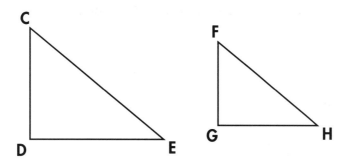

1. Name the remaining pair of corresponding segments.

_____

2. Name all pairs of corresponding angles.

_____

3. Identify the two similar triangles to the right.
Then name all corresponding parts.

_____

_____

_____

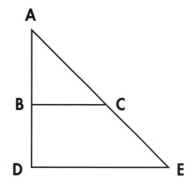

4. Name all corresponding parts of
ΔXYZ and ΔMNO.

_____

_____

_____

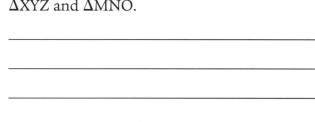

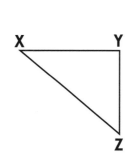

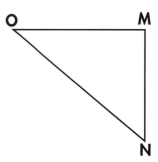

CONGRUENCY AND SIMILAR POLYGONS

Similar polygons are the same shape but different sizes. Exactly how closely related are the corresponding parts of similar polygons? An important rule to remember is that the measurements of corresponding angles are always equal.

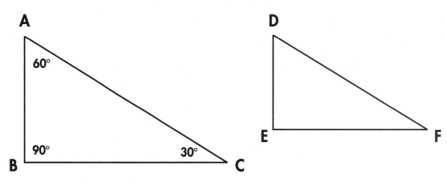

1. Name the three pairs of corresponding angles in the similar triangles above.

_____

2. Keeping in mind the rule about corresponding angles, what is the measure of

   a. angle DEF? _____

   b. angle EFD? _____

   c. angle FDE? _____

3. Identify the similar rectangles in the figure below. Name the similar segments, similar angles, and the measure of each angle.

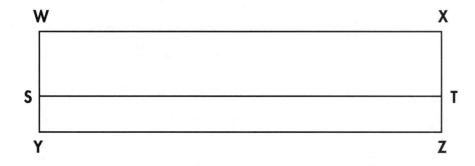

_____

_____

_____

PROPORTIONALITY IN SIMILAR FIGURES

Corresponding angles of similar polygons are equal. But how do the sides of similar polygons relate? Since the corresponding angles are the same, the size of each segment is the only difference between them. Therefore, the lengths of corresponding sides of similar figures are proportional. This means that the lengths of all segments differ by the same proportion. Look at the example below.

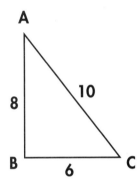

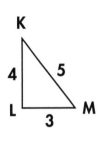

To find out the proportionality, or constant of similarity, for the corresponding sides, we write them as fractions. Since KL corresponds to AB, we can say that they are proportional according to the ratio of the lengths of their sides: $\frac{KL}{AB} = \frac{4}{8}$.

The ratios of the other corresponding sides are as follows:
LM corresponds to BC $= \frac{LM}{BC} = \frac{3}{6}$. MK corresponds to CA $= \frac{MK}{CA} = \frac{5}{10}$.

Notice that all these fractions can be reduced to $\frac{1}{2}$. You now know exactly how much smaller triangle KLM is compared to triangle ABC.

The constant of similarity is $\frac{1}{2}$.

What is the constant of similarity for the figures below, comparing small to large?

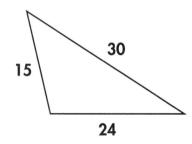

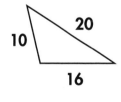

CONSTANT OF SIMILARITY IN POLYGONS

Finding the constant of similarity only requires the use of one corresponding pair of segments in a similar polygon. The constant also depends upon in what order you compare similar figures—large to small or small to large.

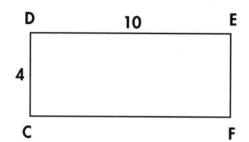 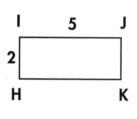

Figures CDEF and HIJK are similar. Find the constant of similarity comparing HIJK to CDEF. For example, choose one pair of corresponding sides. Write them in the same order in which you are comparing figures: HI corresponds to CD, or $\frac{HI}{CD} = \frac{2}{4} = \frac{1}{2}$.

So the constant of similarity between these figures, comparing small to large, is $\frac{1}{2}$.

What about the opposite instance?

1. Find the constant of similarity between the figures above, comparing CDEF to HIJK.

_____

2. Find the constant of similarity of ABDC to WXZY.

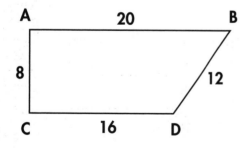 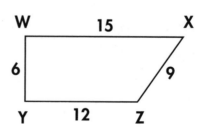

_____

3. Find the constant of similarity of WXZY to ABDC.

_____

MORE CONSTANT OF SIMILARITY

The constant of similarity can be used as a tool to determine the lengths of the unknown sides of polygons. Look at the example below.

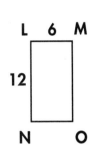

1. What is the constant of similarity in comparing figure LMON to QRTS?

Choosing one pair of corresponding sides, $\frac{LN}{QS} = \frac{12}{18} = \frac{2}{3}$ is the constant of similarity.

You should have noticed that segment LM is 6, but the length of corresponding segment QR is not given. We can determine the length of the unknown segment by using the constant of similarity as described below.

Let's call the length of QR "x."

Since the constant of similarity is the same for all corresponding pairs, $\frac{2}{3} = \frac{LM}{x}$ allows us to find the length of QR.

Substitute the value of LM. $\frac{2}{3} = \frac{6}{x}$

Simplify. $2x = 6\,(3)$
$2x = 18$

Solve for x. $x = 9$
$QR = 9$

1. Using the method described above, determine the length of segment OP.

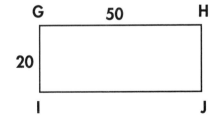
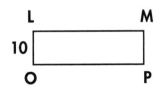

2. Determine the length of segment AB.

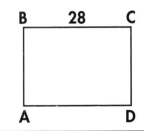
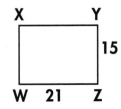

introduction to geometry

PROPORTIONALITY OF POLYGONS

Use the constant of similarity to find the length of the unknown side in each polygon below. Assume that each pair of polygons is similar. Show your work.

---

1. Find the length of segment TU.

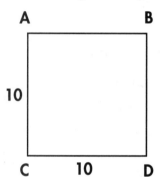 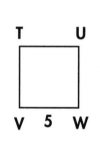

3. Find the length of segment US.

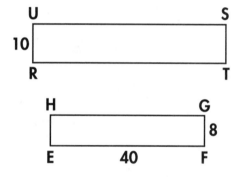

---

2. Find the length of segment LM.

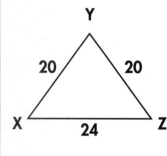 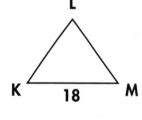

4. Find the length of segment OQ.

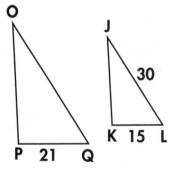

---

everyday

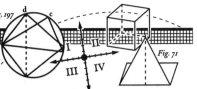

introduction to geometry

MORE POLYGON PROPORTIONS

Use the constant of similarity to determine the lengths of the missing segments in the figures below. Assume that each pair of figures is similar. Show your work.

1. Find the length of segment DF.

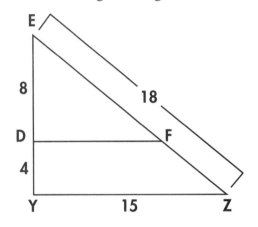

4. Find the length of segment AB.

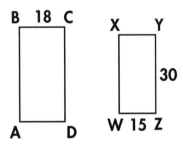

2. Find the length of segment RS.

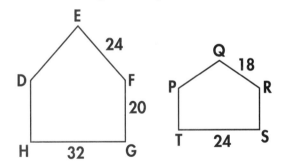

5. Find the length of segment ST.

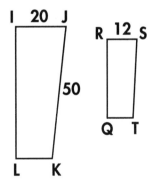

3. Find the length of segment EF.

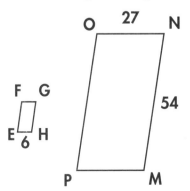

6. Find the length of segment UW.

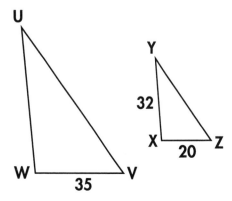

e v e r y d a y | **introduction to geometry**

SIMILAR FIGURES REVIEW

1. **Circle the similar figures. Write two reasons why they are similar. Write why the remaining figure is not similar.**

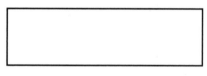

_____

_____

_____

_____

2. **Identify each pair of corresponding angles and sides in the similar figures below.**

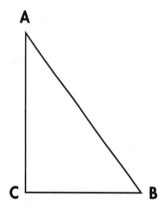

_____

_____

_____

VOLUME OF A CYLINDER

Remember that a cylinder is a top and bottom made of circles with a rectangle wrapped around the outside. The formula for volume is V = area of the base × height.

Since the area of the circular base is found by the formula A = πr^2, the overall formula for finding the volume of a cylinder is V = $\pi r^2 h$.

Find the volumes of the cylinders below.

1. V = _____

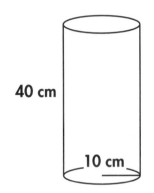

40 cm

10 cm

2. V = _____

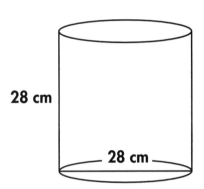

28 cm

28 cm

3. V = _____

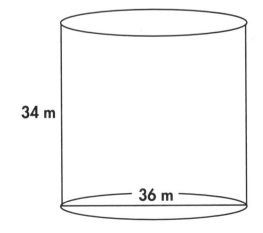

34 m

36 m

4. V = _____

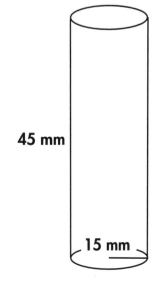

45 mm

15 mm

FS-10607 Everyday Introduction to Geometry

everyday **introduction to geometry**

AREA OF A SODA CAN

To do this activity, you will need a soda can. You will also need a metric ruler. Be sure to record all measurements to the nearest tenth.

1. Measure the height and diameter of the soda can. Record these measurements and draw your can below.

2. Using the measurements above, determine the volume of your soda can. Show your work.

    V = _____

3. How much liquid does your can hold in milliliters? Look at the label, and write the amount of liquid it can hold in ml. How does this number compare with the volume you calculated above? (**Hint:** One cm^3 is the same as one milliliter.)

    _____

    _____

    _____

    _____

TOTAL AREA OF A PRISM

Cut out the shapes below around the outside edges only. Determine the total area of each figure in centimeters. Remember that total is found by adding together the area of all the faces.

Figure 1 _____

Figure 2 _____

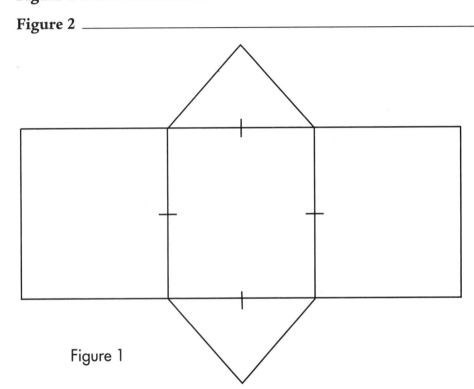

Figure 1

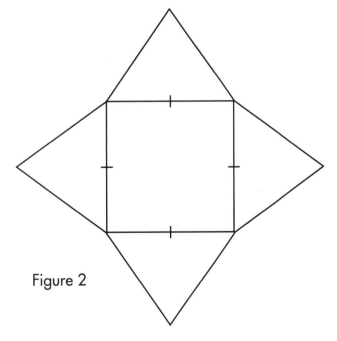

Figure 2

FS-10607 Everyday Introduction to Geometry

introduction to geometry

AREA OF A TRAPEZOID

A trapezoid is a quadrilateral with one pair of parallel sides. To determine the area of a trapezoid, the following formula is used: $A = \frac{1}{2} h (b_1 + b_2)$.

Look at the trapezoid below. It has two bases, b_1 and b_2. The height is given from one vertex to the opposite side.

Find the area of the trapezoid below.

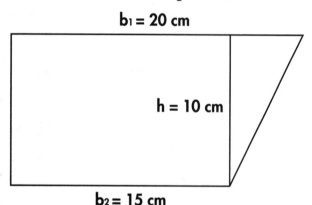

b_1 = 20 cm

h = 10 cm

b_2 = 15 cm

Example: $A = \dfrac{10\,(20 + 15)}{2}$

$A = \dfrac{10\,(35)}{2}$

The area of the trapezoid is 175 cm².

Find the areas of the trapezoids below. Show your work.

1. A = _____

18 cm

6 cm

24 cm

2. A = _____

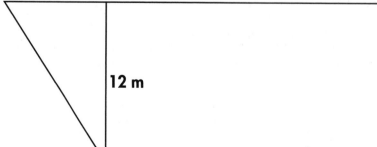

38 m

12 m

26 m

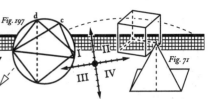

everyday

introduction to geometry

TRAPEZOID PRACTICE

Find the area of each trapezoid below.

1. A = _____

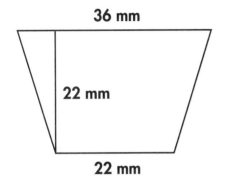

36 mm

22 mm

22 mm

4. A = _____

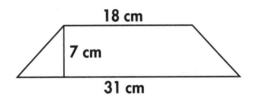

18 cm

7 cm

31 cm

2. A = _____

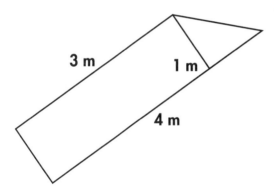

3 m

1 m

4 m

5. A = _____

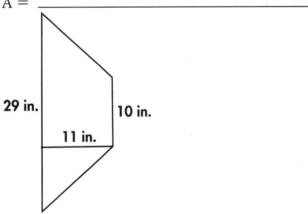

29 in.

10 in.

11 in.

3. A = _____

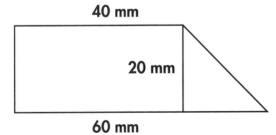

40 mm

20 mm

60 mm

6. A = _____

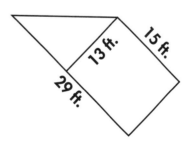

13 ft.

15 ft.

29 ft.

Name_____ Date_____

CLASSIFYING TRIANGLES BY SIDES

Triangles are classified in two ways: (1) according to their angles and (2) according to their sides. The three examples below are classified according to their sides. Count the number of sides in each triangle that are congruent, or the same length.

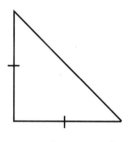

Isosceles triangle—
when two sides are congruent

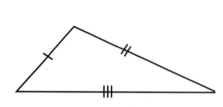

Scalene triangle—
when no sides are congruent

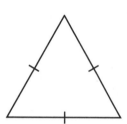

Equilateral triangle—
when all three sides are congruent

Identify the triangles below as scalene, equilateral, or isosceles. Explain why you classified each triangle this way.

1.

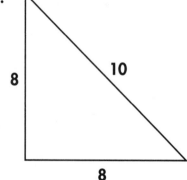

2.

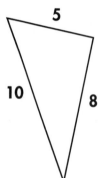

3.

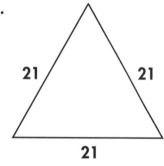

_____

_____

4. Draw and label an equilateral triangle.

5. Draw and label an isosceles triangle.

6. Draw and label a scalene triangle.

CLASSIFYING TRIANGLES BY ANGLES

Similar rules for the lengths of the sides of a triangle apply to the measurements of angles in a triangle. A triangle is identified as an acute triangle if all the angles within it are acute (less than 90°). A triangle is called a right triangle if it contains a right angle. Obtuse triangles contain an obtuse angle (one that is greater than 90°). Equiangular triangles are those with all the same angles.

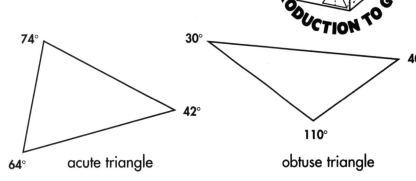

acute triangle

obtuse triangle

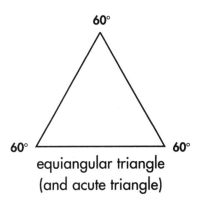

equiangular triangle
(and acute triangle)

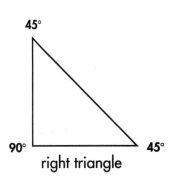

right triangle

Classify the triangles below by their angles.

1. _____

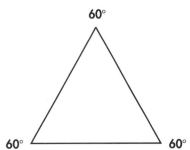

2. _____

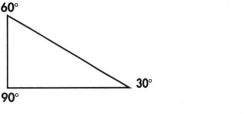

3. _____

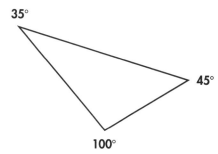

4. _____

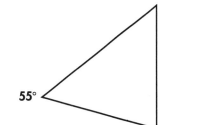

RATIOS OF AREAS OF SIMILAR POLYGONS

Similar polygons have corresponding sides, perimeters, and areas. In this activity, the ratios of the areas of the polygons will be found and compared to the constant of similarity.

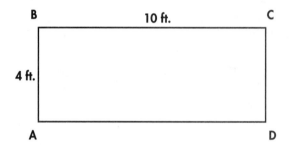

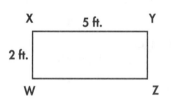

1. What is the constant of similarity between the figures ABCD and WXYZ?

2. Find the area of figure WXYZ. Show your work.

   A = _____

3. Find the area of figure ABCD. Show your work.

   A = _____

4. What is the ratio of the two areas? (**Hint:** Divide the area of WXYZ by ABCD.)

5. How does this ratio compare to the constant of similarity?

PERIMETER AND CONSTANT OF SIMILARITY

You have seen that finding the constant of similarity helps you understand the relationship between two similar figures. In this activity, the perimeter of two similar figures will be compared.

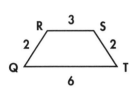

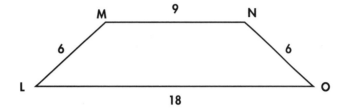

1. Find the constant of similarity between figures QRST and LMNO. Show your work.

2. Find the perimeter of figure QRST.

P = _____

3. Find the perimeter of figure LMNO.

P = _____

THERE'S NOT GONNA BE, LIKE, A QUIZ ON THIS...RIGHT?

4. Find the ratio of the perimeters QRST to LMNO. How does this compare with the constant of similarity?

introduction to geometry

WORKING WITH SCALE

Scale models are small versions of real cars, trains, airplanes, and other objects in the real world that are shrunk down to "scale" size. They can even be thought of as similar figures. To build a realistic scale model, the full-sized item must be carefully examined, and a scale for the model should be decided upon. For example, suppose you wanted to build a model of a boxcar that is 40 feet long, 20 feet high, and 10 feet wide. If you wanted to build a $\frac{1}{10}$ scale model, how big would it be?

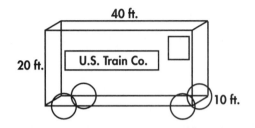

Example:		Original	Model
Length	40	$40 \times \frac{1}{10} = 4$ ft.
Width	10	$10 \times \frac{1}{10} = 1$ ft.
Height	20	$20 \times \frac{1}{10} = 2$ ft.

So the model would have to be 4 feet long, 2 feet high, and 1 foot wide.

Using the original train above, decide on a scale for a model that could fit on a table in your home. Using this scale, determine the dimensions of the model. Then draw the model as it would appear in your home.

everyday

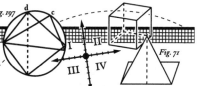

introduction to geometry

SCALE MODEL AND DRAWINGS

1. The figure below needs to be scaled down to a $\frac{1}{4}$ scale. Using a ruler, first measure the dimensions of all sides. Then determine the new dimensions and redraw the figure using a ruler.

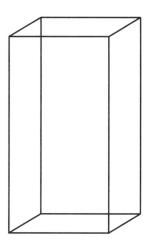

2. Scale this figure to a $\frac{1}{2}$ scale drawing. Remember, you need to measure all three dimensions.

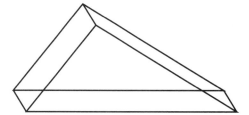

everyday **introduction to geometry**

SCALE DRAWINGS

Similar figures and scale are used in plans for buildings and maps. In this activity, the actual dimensions of a house will be determined by examining the scale drawings of the house plans.

Suppose a large box is going to be built. Every centimeter of the box plans below equals 2 meters. How are the dimensions of the box determined?

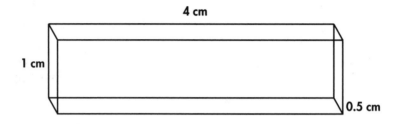

4 cm

1 cm

0.5 cm

	Scale Drawing	Box
Length	4 cm	4 × 2 = 8 meters
Width	0.5 cm	0.5 × 2 = 1 meter
Height	1 cm	1 × 2 = 2 meters

1. If the scale on the box above is 1 cm = 6 feet, what would be the dimensions of the box in feet?

_____

2. Suppose the scale on the figure below is 1 centimeter = 1 foot. Using a ruler, measure the figure and determine what the dimensions of the real object would be.

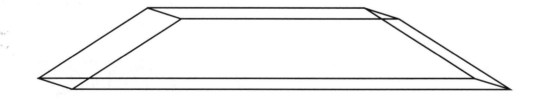

_____

HOUSE PLANS

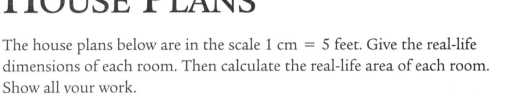

The house plans below are in the scale 1 cm = 5 feet. Give the real-life dimensions of each room. Then calculate the real-life area of each room. Show all your work.

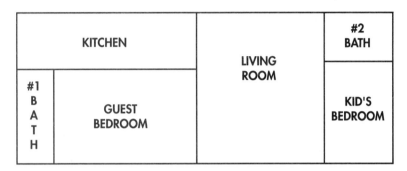

| KITCHEN | LIVING ROOM | #2 BATH |
| #1 BATH | GUEST BEDROOM | | KID'S BEDROOM |

1. Kitchen

_____

2. Bathroom #1

_____

3. Guest Bedroom

_____

4. Living Room

_____

5. Kid's Bedroom

_____

6. Bathroom #2

_____

DUDE...THINK YOU COULD, LIKE, HURRY IT UP WITH THE, LIKE, BATHROOM..?

COLOR THE CUBE

Small cubes have been stacked and glued together to form this larger cube below.
Use the cube to answer the questions.

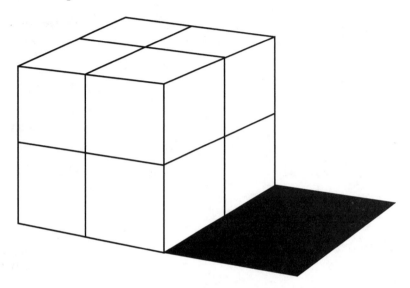

1. How many small cubes were used? _____

2. If the cube was dropped in a pail of paint:

   a. how many small cubes would have paint on three sides? _____

   b. how many small cubes would have paint only on two sides? _____

   c. how many small cubes would have paint only on one side? _____

   d. how many small cubes would not have paint on any side? _____

3. Answer questions 1 and 2 for a 3 × 3 cube.   1. _____   2. a. _____

                                                                        b. _____

                                                                        c. _____

                                                                        d. _____

4. Answer questions 1 and 2 for a 4 × 4 cube.   1. _____   2. a. _____

                                                                        b. _____

                                                                        c. _____

                                                                        d. _____

Answer Key

Page 1

1.–4.

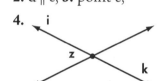

A   B   C

5. Answers will vary. Possible answers include: **a.** segment AC has endpoints, **b.** line AC has no midpoint.

Page 2

1. no, no angles; **2.** no, no angles; **3.** yes; **4.** yes; **5.** no, open; **6.** yes; **7.** yes; **8.** no, no vertices; **9.** no, no angles

Page 3

1. d ⟷

   e ⟷

2. d ∥ e; **3.** point e;

4.

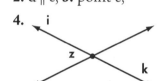

Page 4

1. ∠ACG; **2.** ∠HKP; **3.** ∠ZXY; **4.** ∠WQS; **5.** ∠TMB; **6.** ∠STW;

7.

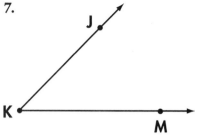

8.

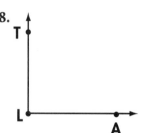

9.

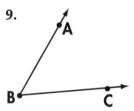

Page 5

1. ∠DAB, ∠ABC, ∠BCD, ∠CDA, vertices A, B, C, D;
2. ∠SQR, ∠QRT, ∠RTU, ∠TUS, ∠USQ, vertices Q, R, T, U, S;
3. ∠XWY, ∠WYX, ∠YXW, vertices X, W, Y; **4.** ∠OLM, ∠LMN, ∠MNO, ∠NOL, vertices L, M, N, O; **5.** ∠STU, ∠TUV, ∠UVS, ∠VST, vertices S, T, U, V;
6. Answers will vary.

Page 6

Number of Sides	Greek Word	Name
3	—	triangle
4	—	quadrilateral
5	penta	pentagon
6	hexa	hexagon
7	hepta	heptagon
8	octa	octagon
9	nona	nonagon
10	deca	decagon

1. quadrilateral; **2.** hexagon; **3.** octagon; **4.** triangle; **5.** nonagon; **6.** pentagon; **7.** decagon; **8.** heptagon; **9.** quadrilateral

Page 7

1. a. parallelogram, rectangle; **b.** parallelogram, rhombus, square, rectangle; **c.** parallelogram, rhombus; **d.** parallelogram, rectangle; **e.** trapezoid; **f.** trapezoid; **2.** Figures will vary.

Page 8

Answers will vary.

Page 9

Answers will vary according to tile pattern created.

Page 10

1. quadrilateral (rectangle); **2.** hexagon; **3.** pentagon; **4.** not a polygon; **5.** quadrilateral (rectangle); **6.** quadrilateral (trapezoid); **7.** octagon; **8.** not a polygon; **9.** quadrilateral (rhombus); **10.** quadrilateral (square); **11.** not a polygon; **12.** hexagon; **13.** triangle; **14.** nonagon; **15.** heptagon

Page 11

1. 14 cm; **2.** 21 cm; **3.** 6 cm; **4.** 68 cm; **5.** 15 cm; **6.** 22 cm; **7.** 60 cm; **8.** 17 cm; **9.** 12 cm

Page 12

1. 9 cm; **2.** 15 cm; **3.** 13 cm; **4.** 9.8 cm; **5.** 6 cm; **6.** 14.8 cm; **7.** 13.7 cm; **8.** 10 cm

Page 13

1. $a = 4$ m, $d = 4$ m, $c = 4$ m; **2.** $a = 6$ mm, $b = 6$ mm, $c = 6$ mm, $d = 6$ mm, $e = 6$ mm; **3.** $y = 7$ cm, $x = 2$ cm, $z = 2$ cm; **4.** $a = 18$ m, $c = 18$ m

Page 14

1. A; **2.** A, B; **3.** 21 mm;
4. Drawings will vary;
5. Drawings will vary; **6.** Z;
7. Drawings will vary; **8.** 22 mm

Page 15

1. 5 cm, 5 cm; **2.** 12 cm, 6 cm,
6 cm; **3.** 12 cm; **4.** an infinite
number

Page 16

1. KU, OG, MT, UX, XS; **2.** KW,
OW, GW; **3.** OG

Page 17

1. 90 degrees; **2.** 90 degrees;
3. four; **4.** four; **5.** 360 degrees

Page 18

1. ∠CAE, ∠EAD, ∠DAB;
2. ∠CAD; **3.** 55 degrees

Page 19

1. ∠DBC, ∠DBF; **2.** ∠DBE,
∠DBA; **3.** ∠EBC, ∠CBF, ∠ABF,
∠ABE; **4.–6.** Answers will vary.

Page 20

1. 60 degrees; **2.** 30 degrees;
3. 60 degrees; **4.** ∠WYZ, ∠ZYR,
∠RYX, ∠XYW; **5.** ∠WYT, ∠TYZ,
∠ZYH, ∠HYR; **6.** ∠XYT, ∠WYH,
∠TYR; **7.** 90 degrees;
8. 45 degrees

Page 21

1. Cool, wet climate because roof
is at a steep pitch; **2.–3.** Answers
will vary.

Page 22

1. 90 degrees; **2.** 90 degrees;
3. No, because parallel lines
never meet; **4.** Answers will vary.

Page 23

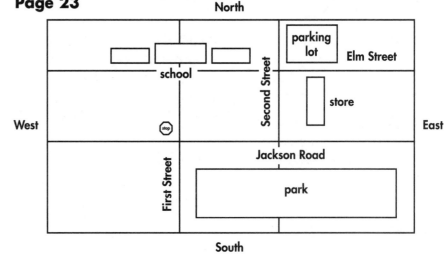

Page 24

1. 31.4 cm; **2.** d = 10 cm,
c = 31.4 cm; **3.** 28.26 cm

Page 25

1. 18.84 cm; **2.** 3 cm; **3.** 8 cm;
4. 25.12 cm; **5.** 24 cm; **6.** 75.36
cm; **7.** 8.5 cm; **8.** 53.38 cm

Page 26

1. 15.92; **2.** 23.89; **3.** 15.92;
4. 138.16; **5.** 13.54; **6.** 4.62

Page 27

1. 300 yards; **2.** 5000 yds.²;
3. inches²; **4.** meters²; **5.** 22 feet;
6. 24 feet²

Page 28

1. 42 m²; **2.** 24 m²; **3.** 66 m²;
4. P = 42 m, 24 less than area;
5. Area is the entire surface to
be covered, perimeter is only
outside.

Page 29

1. 16 m; **2.** 12 m²; **3.** 16 cm²;
4. Answers will vary.

Page 30

1. 4 cm; **2.** 4 cm; **3.** 12 cm

Page 31

1. Television: 528 in.²; VCR:
160 in.²; Receiver: 216 in.²;
CD Player: 196 in.²;
2. Answers will vary.

Page 32

1.

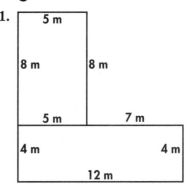

2. 40 m²; **3.** 48 m²; **4.** 88 m²;
5. Smaller rectangles may vary.
One possible solution is below.

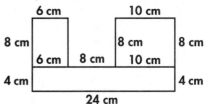

6. 48 cm², 80 cm², 96 cm²;
7. 224 cm²

Page 33

1. 30 m²; **2.** A = 18 m², P = 20 m;
3. A = 126 cm², P = 54 cm

Page 34

1. A = 48 cm², P = 40 cm;
2. A = 165 in.², P = 54 in.;
3. A = 44 ft.², P = 48 ft.;
4. A = 612 m², P = 114 m

Page 35

1. 40 cm²; **2.** two triangles,
20 cm²; **3.** 20 cm²; **4.** 30 cm²;
5. 4 ft.²

Page 36

1. 171 cm²; **2.** 11 ft.²; **3.** 221 in.²;
4. 522 m²; **5.** 54 mm²

Page 37

1. 297 in²; **2.** 960 cm²; **3.** 806 mm²;
4. 495 cm²; **5.** 720 mm²

Page 38

1. b = 8 cm; **2.** h = 6 in.;
3. h = 10 mm; **4.** b = 12 ft.

Page 39

1. a. 128 cm², **b.** 48 cm, **c.** 88 cm²

Page 40

1. 839 ft.²; **2.** 550 ft.²

Page 41

1. 113.04 cm²; **2.** 254.34 in.²

Page 42

1. 4298.66 mm²; **2.** 2289.06 mm²;
3. 254.34 m²; **4.** 452.16 cm²

Page 43

Area per pie = 78.5 in.²/12 in.
= 6.54 ft.²; 6.54 × 21 pies =
137.34 ft.²; oven area = 96 ft.²;
No, there is not enough room.

Page 44

1. r = 3.5 cm, d = 7 cm;
2. r = 5.2 cm, d = 10.4 cm;
3. r = 2.5 ft., d = 5 ft.;
4. r = 3.7 m, d = 7.4 m;
5. r = 2.9 in., d = 5.8 in.;
6. r = 4.2 mm, d = 8.4 mm

Page 45

1. 90 in.²; **2.** 122 ft.²; **3.** 384 cm²;
4. 396 mm²

Page 46

Answers will vary.

Page 47

1. 175.84 cm²; **2.** 395.64 mm²

Page 48

1. 30 liters; **2.** 100 liters;
3. 1.5 liters; **4.** 1800 cm³; **5.** 64 ft.³

Page 49

1. SA = 56 m², V = 24 m³;
2. SA = 822 m², V = 1496 m³;
3. SA = 22 m², V = 5 m³

Page 50

Answers will vary.

Page 51

1. must be same shape, must be
different size;

2.

3.

4.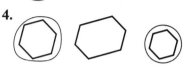

5.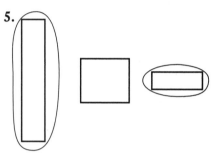

6.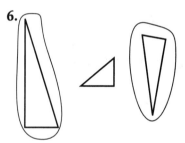

Page 52

1. Same parts, only sizes differ;
2. wheels, bumpers, windshield,
etc.; **3.** ∠B and ∠R, ∠A and ∠S;
4. ∠W and ∠H, ∠X and ∠I,
∠Y and ∠J

Page 53

1. ABCD and LMON are similar.
∠A and ∠L, ∠B and ∠M, ∠C
and ∠N, ∠D and ∠O; **2.** ∆DFE
and ∆GIH are similar. ∠D and
∠G, ∠E and ∠H, ∠F and ∠I;
3. WXZY and ABDC are similar.
∠W and ∠A, ∠X and ∠B, ∠Y
and ∠C, ∠Z and ∠D

Page 54

1. CE and FH; **2.** ∠C and ∠F, ∠D and ∠G, ∠E and ∠H; **3.** ΔADE, ΔABC; ∠A and ∠A, ∠B and ∠D, ∠C and ∠E; AB and AD, AC and AE, BC and DE **4.** ∠Y and ∠M, ∠X and ∠O, ∠Z and ∠N, YX and MO, YZ and MN, XZ and ON

Page 55

1. ∠A and ∠D, ∠B and ∠E, ∠C and ∠F; **2. a.** 90°, **b.** 30°, **c.** 60°; **3.** WXZY and WXTS are similar. WX and WX, XZ and XT, ZY and TS, YW and SW; ∠W and ∠W, ∠X and ∠X, ∠Z and ∠T, ∠Y and ∠S; all angles = 90°

Page 56

$\frac{2}{3}$

Page 57

1. 2; **2.** $\frac{4}{3}$; **3.** $\frac{3}{4}$

Page 58

1. 25; **2.** 20

Page 59

1. 5; **2.** 15; **3.** 50; **4.** 42

Page 60

1. 10; **2.** 15; **3.** 12; **4.** 36; **5.** 30; **6.** 56

Page 61

1.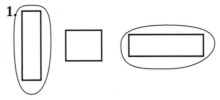

reasons: same shape, different size, congruent parts, same angles; **2.** ∠A and ∠R, ∠B and ∠S, ∠C and ∠T; CA and TR, AB and RS, BC and ST

Page 62

1. 12,560 cm³; **2.** 17,232.32 cm³; **3.** 34,590.24 m³; **4.** 31,792.5 mm³

Page 63

Answers will vary.

Page 64

figure 1 = 66 cm²; figure 2 = 33.25 cm²

Page 65

1. 126 cm²; **2.** 384 m²

Page 66

1. 638 mm²; **2.** 3.5 m²; **3.** 1000 mm²; **4.** 171.5 cm²; **5.** 214.5 in.²; **6.** 286 ft.²

Page 67

1. isoceles; two congruent sides; **2.** scalene; no congruent sides; **3.** equilateral; all congruent sides; **4.–6.** Drawings will vary.

Page 68

1. equiangular, acute; **2.** right; **3.** obtuse; **4.** acute

Page 69

1. 2; **2.** 10 ft.²; **3.** 40 ft.²; **4.** 1:4 or $\frac{1}{4}$; **5.** twice, or squared amount

Page 70

1. $\frac{1}{3}$; **2.** 13; **3.** 39; **4.** $\frac{13}{39} = \frac{1}{3}$, same as constant of similarity

Page 71

Answers will vary.

Page 72

All drawings should be identical to the original except they should be drawn to scale.

Page 73

1. 6 ft. × 24 ft. × 3 ft.; **2.**

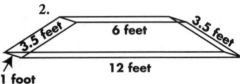

Page 74

1. 24.5 × 7.5 = 183.75 ft.²; **2.** 12.5 × 5 = 62.5 ft.²; **3.** 19.5 × 12.5 = 243.75 ft.²; **4.** 17.5 × 20 = 350 ft.²; **5.** 9.5 × 13.5 = 128.25 ft.²; **6.** 9.5 × 6.5 = 61.75 ft.²

Page 75

1. 8; **2. a.** 8; **b.** 0; **c.** 0; **d.** 0; **3. (1)** 27; **(2) a.** 8; **b.** 12; **c.** 6; **d.** 1; **4. (1)** 64; **(2) a.** 8; **b.** 24; **c.** 24; **d.** 8

FS-10607 Everyday Introduction to Geometry